All About Deer

MICHÈLE DUFRESNE

TABLE OF CONTENTS

What Is a Deer?	2
Moose	4
Elk	6
White-Tailed Deer	8
Reindeer	10
Antlers	12
Habitat	16
Baby Deer	18
Glossary/Index	20

PIONEER VALLEY EDUCATIONAL PRESS, INC

WHAT IS A DEER?

Have you ever seen a deer?

A deer is a large mammal with antlers.

There are many kinds of deer.

They live in most parts of the world.

They live in warm places,

like the rain forests,

and in cold places up north.

elk

white-tailed deer

mule deer

moose

reindeer

black-tailed deer

3

MOOSE

Moose are the largest member of the deer family. They have **massive** bodies and thin legs. They have a short tail and a hump on their shoulders. They have large ears that they turn to help them hear better. Moose like to live in cold places.

≫ **Moose can't live in warm places because they can't sweat.**

ELK

Elk are also very large deer. A male elk, with its antlers, can tower at 9 feet tall. During the fall, elk grow a thicker coat of hair. This keeps them warm in the winter. In the early summer, elk rub up against trees to help **shed** the extra hair from their bodies. They **migrate** to the high mountains where they graze on grass and other plants. Mother elk, called cows, give birth to their babies.

MORE TO EXPLORE

A male elk is called a bull. Bulls make a loud sound called **BUGLING**. You can hear them bugling from miles away.

7

WHITE-TAILED DEER

The white-tailed deer is a small deer. The white-tailed deer's tail is white on the underside. When deer smell danger, they raise their tail and run. The white flash of the tail warns other deer of danger.

MORE TO EXPLORE

When there is danger, the deer will breathe very hard. This is called blowing. Then they run away. When they blow, it lets other deer know about the danger.

You are most likely to see white-tailed deer early in the morning or in the late afternoon when they are looking for food. During the hot summer, they **inhabit** fields and meadows. In the winter, they look for shelter in the forest.

REINDEER

Reindeer are also called caribou. They live in the north where it is very cold. They are strong runners. Even a one-day-old baby reindeer is very fast. Reindeer are good swimmers, too. They can swim across wide rivers and big lakes.

In some places, reindeer are used to pull sleds.

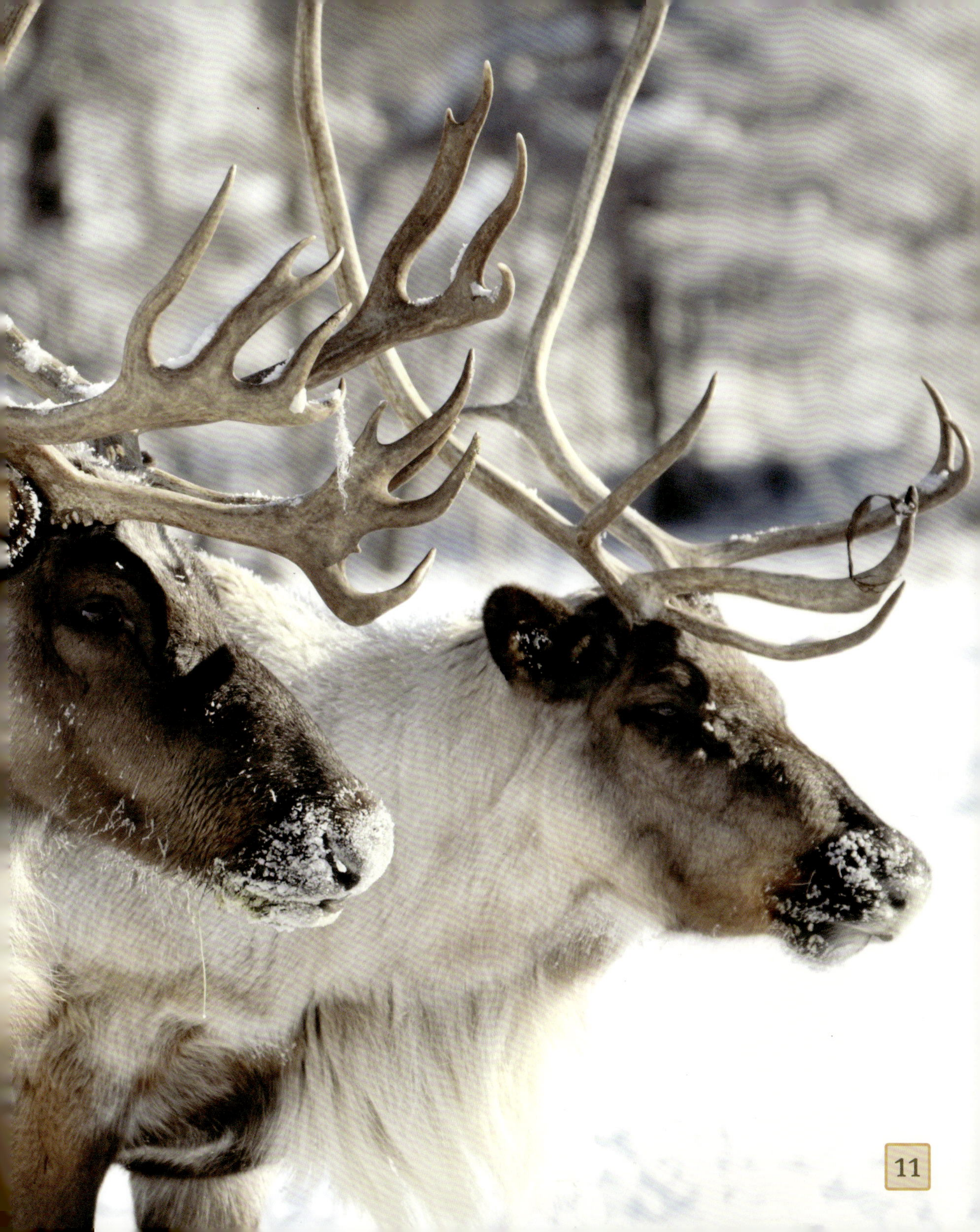

ANTLERS

Most male deer are called bucks.
Bucks have antlers.
Their antlers begin to grow
in the spring.
Antlers can grow very fast.
They can grow up to one quarter
of an inch each day.
The antlers fall off, or shed, each winter.
The next spring, new antlers
begin to grow again.

13

During **mating** season, bucks will use their antlers to fight each other. The two bucks circle each other. They bend their back legs, lower their heads, and charge. They lock their antlers together to see who is the strongest. The female deer likes the strongest buck.

More to Explore

Each year, the buck's antlers will grow bigger and stronger. A deer has to have enough food to eat to grow antlers. When there are too many deer and not enough food, bucks will have smaller antlers.

>>> **The only female deer that has antlers is the reindeer.**

HABITAT

Some deer like to live in fields near the forest. They like to live where they can eat plants and grass and still be close to trees. They want to be able to run for cover if there is danger.

There are deer that live in the mountains. When cold weather comes, they will move down off the mountain to stay warm. When spring comes, they go back up to the mountains.

BABY DEER

A female deer is called a doe.

A doe will have between

one and three babies at a time.

The babies are often born

in May or June.

The young deer is usually called a **fawn**.

When a fawn sees danger,

it will lie down in the grass

and stay very still.

A doe and her fawns live in a family group until the fawns stop drinking their mother's milk.

MORE TO EXPLORE

19

GLOSSARY

bugling
a loud call of the male deer, elk, or moose

fawn
a young deer

inhabit
to live in a place

massive
very large and heavy

mating
when a male and female animal come together to make babies

migrate
to move from one area to another at different times of the year

shed
to cast off a natural covering, such as fur

INDEX

antlers 12-13, 14-15
black-tailed deer 3
bucks 12-13
bugling 7
bull 7
caribou 10-11
danger 8, 16, 18
doe 18-19
elk 2, 6-7
fawn 18-19
forest 16
habitat 16
mating 14
moose 3, 4-5
mule deer 3
reindeer 3, 10-11, 15
shed 6, 12
swim 10
rain forest 2
white-tailed deer 2, 8-9